There is a reference page ON THE BACK COVER OF THIS BOOK and as a removable sheet in the back of this book.

Thank you for buying this book!
It is proof that a person can use factoring later in their life.

It also keeps me in the lap of luxury lol

My goal is for you to have a thorough understanding of factoring by the time you complete this workbook. Answers are included in the back of the book and I have created a YouTube playlist working through select problems so you can learn step-by-step how to factor multiple types of polynomials.

If you still have questions, **send me a DM @bestmathtutor**.

If you already love math, FANTASTIC!
If you don't, try one or more of the following:

a. be positive
b. stop whining
c. smile more
d. be awesome

This will lead to sweet grades, college, travel, fast cars, a job you'll enjoy, and as much avocado as you want!

Speaking from experience,

ISBN: 978-1-941775-98-1

Ordering Information:
Quantity sales. Special discounts are available on quantity purchases by corporations, associations, and others. For bulk discounts or orders by U.S. trade bookstores and wholesalers, contact the publisher at the phone number or email above.

Notes to study b4

Put all of the important stuff to reme

Factoring

Numbers have **factors**.

A factor pair is a set of **2 terms** that **multiply** to get another term.

> **Another way to think of factors:**
> Factors are terms that, when multiplied, equal the same term.
> For example, 1 • 12 = 12 and 2 • 6 = 12 and 3 • 4 = 12.
> **1 and 12, 2 and 6, 3 and 4 are all factors of 12.**

FOR EXAMPLE:

$$2 \times 3 = 6 \quad \Rightarrow \quad \text{2 and 3 are factors of 6}$$
$$5 \times 4 = 20 \quad \Rightarrow \quad \text{5 and 4 are factors of 20}$$
$$7 \times 9 = 63 \quad \Rightarrow \quad \text{7 and 9 are factors of 63}$$
$$-3 \times 8 = -24 \quad \Rightarrow \quad \text{-3 and 8 are factors of -24}$$

2 terms   another term

"Factoring" is the process of FINDING THE FACTORS

Some easy factors to start:

6: 1 (6), 2 (3)

20: 1 (20), 2 (10), 4 (5)

63: 1 (63), 7 (9)

Graphing Calculator FACTOR TIP

Step 1: Go to Y=

Step 2: Enter the number you want factors of divided by x.
   For example:
   y = 45/x
   y = -20/x
   y = 250/x

Step 3: Go to TABLE and the factors of that number will be listed as x and y ⇨

Factors of 250 using the phone app:

Graphing Calculator Plus

(available in the App Store)

ΔTbl=1	Y₁
0	N/A
1	250
2	125
3	83.33333333
4	62.5
5	50
6	41.66666667

| STAT Plot | TBLSET | Format | Redo | Table |
| Y= | Window | < Fx > | Undo | Graph |

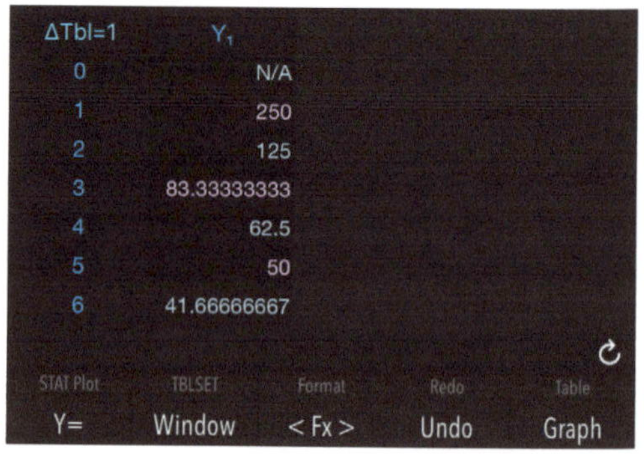

Factors

Find the factors of:

1. 30 ______
2. 45 ______
3. 6 ______
4. 8 ______
5. 19 ______
6. 25 ______
7. 64 ______
8. 100 ______
9. -8 ______
10. -20 ______
11. -12 ______
12. -50 ______
13. -80 ______
14. -6 ______
15. -25 ______

Greatest Common Factor between 2 terms

Find the GCF of:

16 and 18 (2)
16: 1·16, 2·8, 4·4
18: 1·18, 2·9, 3·6

20 and 24 (4)
20: 1·20, 2·10, 4·5
24: 1·24, 2·12, 3·8, 4·6

14x and 21 (7)
14x: {1·14x, 2·7x; 1x·14, 2x·7}
21: 1·21, 3·7

Finding the greatest common factor is simply **finding the biggest number that can be divided from two or more terms**.

For example, find the GCF of **16** and **18**:
So I ask myself, "*what is the biggest number that can be divided out of* **BOTH 16 and 18**?"

I mentally review the factors of **16** and **18**, or take a moment and write them down to visually compare.

After reviewing the factors of 16 and 18, I see that the largest number that can be divided from BOTH 16 and 18 is **2**.
The GCF is 2.

between 3 terms

Find the GCF of:

10, 20 and 25 (5)
10: 1·10, 2·5
20: 1·20, 2·10, 4·5
25: 1·25, 5·5

18, 9 and 21 (3)
18: 1·18, 2·9, 3·6
9: 1·9, 3·3
21: 1·21, 3·7

8x, 12 and 24 (4)
8x: 1·8x, 2·4x; 8x·1, 2x·4
12: 1·12, 2·6, 3·4
24: 1·24, 2·12, 3·8, 4·6

Now find the GCF of **10**, **20** and **25**:
So I ask myself, "*what is the biggest number that can be divided out of ALL THREE NUMBERS* (**10, 20 and 25**)?"

I mentally review the factors of each number and realize that the largest number that can be divided FROM ALL THREE NUMBERS is **5**.
The GCF is 5.

Finding the GCF is simple and YOU CAN DO IT! Try the problems on page 5 and check your answers in the back of the book.

GCF between 2 or 3 terms

Find the GCF PART 1:

1. 4 and 12 ______
2. 9 and 3 ______
3. 8 and 20 ______
4. 16 and 30 ______
5. 14, 21 and 28 ______
6. 18, 3 and 27 ______
7. 64, 24, and 8 ______
8. 2x and 10 ______
9. 4x and 8 ______
10. 3x and 12 ______
11. 14x and 21 ______
12. 20x, 30 and 90 ______
13. 15x, 5 and 40 ______
14. 8x, 16 and 32 ______
15. 11x, 22 and 99 ______

Find the GCF PART 2:

1. x^2 and xy ______
2. $6a^3$ and $3a^2$ ______
3. $3x$ and $9y$ ______
4. $20c^3$ and $15cd^3$ ______
5. $2x^{12}$ and $2x^{11}$ ______
6. $8a^5$ and $17a^3b^2$ ______
7. $14d^2e^4f^2$ and $7d^2e^3f$ ______
8. $5x^{20}$ and $10x^{19}$ ______
9. $6x^3$, $3x^2$ and $9x$ ______
10. $3y^5$, $18y^4$ and $12y^8$ ______
11. $25a^3$, $15ab$ and $18a^2$ ______
12. x^4, $4x^5$ and $8x^3$ ______
13. $15x^3$, $5x^2$ and $20x$ ______
14. $8ab^4c^3$, $16a^3bc^3$ and $32a^2bc^3$ ______
15. $11x^3$, $22x^2$ and $99x$ ______

GREATEST COMMON FACTOR WITH BINOMIALS (2 TERMS)

GCF Greatest Common Factor

~ finding the largest factor common to all terms

① ②
$6x + 2$

Term 1 = 6x Factors of 6x: $1x \cdot 6$, $2x \cdot 3$, $1 \cdot 6x$, ②$\cdot 3x$

Term 2 = 2 Factors of 2: $1 \cdot$ ②

GCF = ②

divide ② from both terms

→ or what do you multiply by 2 to get each term?

$2 \cdot \underline{3x} = 6x$

$2 \cdot \underline{1} = 2$

$2(3x + 1)$

CHECK: $2 \cdot 3x = 6x$ ✓ $2 \cdot 1 = 2$ ✓

---

① ②
$20x - 5$

Greatest common factor of $20x$ and (-5) = ⑤

divide ⑤ from both terms

→ or what do you multiply by 5 to get each term?

$5 \cdot \underline{4x} = 20x$

$5 \cdot \underline{(-1)} = -5$

$5(4x - 1)$

CHECK: $5 \cdot 4x = 20x$ ✓ $5 \cdot (-1) = -5$ ✓

FACTOR 6x + 2

Term 1: 6x

$1x \cdot 6 = 6x$ $1 \cdot 6x = 6x$

$2x \cdot 3 = 6x$ $\mathbf{2} \cdot 3x = 6x$

Term 2: 2

$1 \cdot \mathbf{2} = 2$

<u>The largest factor that 6x and 2 have in common is</u> **2**.

NEXT,
DIVIDE the greatest common factor from both terms

$6x \div \mathbf{2} = \mathbf{3x}$ $2 \div \mathbf{2} = \mathbf{+1}$ *(positive 1)*

3x and +1 are the new terms that go INSIDE the parentheses.

GCF outside parentheses (**new terms inside**)

2 (3x + 1)

FACTOR 20x -5

Term 1: 20x

$1x \cdot 20 = 20x$ $1 \cdot 20x = 20x$

$2x \cdot 10 = 20x$ $2 \cdot 10x = 20x$

$4x \cdot \mathbf{5} = 20x$ $4 \cdot 5x = 20x$

Term 2: -5

$-1 \cdot \mathbf{5} = -5$

$1 \cdot -5 = -5$

<u>The largest factor that 20x and -5 have in common is</u> **5**.

NEXT,
DIVIDE the greatest common factor from both terms

$20x \div \mathbf{5} = \mathbf{4x}$ $-5 \div \mathbf{5} = \mathbf{-1}$ *(negative 1)*

4x and -1 are the new terms that go INSIDE the parentheses.

5 (4x - 1)

GCF with binomials

Factor by GCF Part 1

1. $4x + 2$ ______
2. $8x - 2$ ______
3. $9x + 30$ ______
4. $15x - 10$ ______
5. $30x + 25$ ______
6. $2x - 12$ ______
7. $21x - 14$ ______
8. $7x + 35$ ______
9. $100x - 45$ ______
10. $45x + 9$ ______
11. $5x - 20$ ______
12. $6x - 20$ ______
13. $27x + 18$ ______
14. $8x + 14$ ______
15. $11x - 33$ ______

Factor by GCF Part 2

1. $9x + 3$ ______
2. $11x - 121$ ______
3. $6x + 30$ ______
4. $12x - 10$ ______
5. $36x + 12$ ______
6. $7x - 14$ ______
7. $28x - 40$ ______
8. $5x + 35$ ______
9. $100x - 18$ ______
10. $40x + 15$ ______
11. $5x - 25$ ______
12. $3x - 21$ ______
13. $32x + 24$ ______
14. $15x + 15$ ______
15. $50x - 25$ ______

EXPONENTS ON VARIABLES • FACTORING WITH DIFFERENT VARIABLES (X AND Y)

$4x^2 - 2x$

GCF of $4x^2$ and $(-2x) = 2x$

$2x \cdot 2x = 4x^2$ ✓ $\quad 2x \cdot (-1) = -2x$ ✓

$2x(2x - 1)$

$$2x(2x-1)$$

We now have an "x" in BOTH TERMS. This means we can divide 2 from both terms (2 is the GCF) and ALSO DIVIDE X from both terms!

Divide each term by **2x**.

$4x^2 \div \mathbf{2x} = \mathbf{2x}$

$-2x \div \mathbf{2x} = \mathbf{-1}$

The **GCF** goes OUTSIDE the parentheses, new terms INSIDE.

Answer: **2x** (**2x - 1**)

Check urself b4 u wreck yourself:

2x • **2x** = $4x^2$ ✓

2x • **-1** = $-2x$ ✓

$7x^2 - 14$

GCF of $7x^2$ and $(-14) = 7$

$7 \cdot x^2 = 7x^2$ ✓ $\quad 7 \cdot (-2) = -14$ ✓

$7(x^2 - 2)$

$$7(x^2-2)$$

Now an "x^2," but no x's in the other term. Therefore, we can ONLY divide each term by the numerical GCF of 7. The two terms do not share an x in common. They can stay friends.

Divide each term by **7**.

$7x^2 \div \mathbf{7} = \mathbf{1x^2}$

$-14 \div \mathbf{7} = \mathbf{-2}$

Answer: **7** (**x^2 - 2**)

Check yourself:

7 • **x^2** = $7x^2$ ✓

7 • **-2** = -14 ✓

$12x^2 + 5x$

GCF of $12x^2$ and $5x = x$

$x \cdot 12x = 12x^2$ ✓ $\quad x \cdot 5 = 5x$ ✓

$x(12x + 5)$

$$x(12x+5)$$

Again, there is an "x" in BOTH TERMS. So, we can divide ONLY AN X from both terms. 12 and 5 do not have any common factors.

Divide each term by **x**.

$12x^2 \div \mathbf{x} = \mathbf{12x}$

$5x \div \mathbf{x} = \mathbf{5}$

The **GCF** goes OUTSIDE the parentheses, new terms INSIDE.

Answer: **x** (**12x** + **5**)

Check yourself:

x • **12x** = $12x^2$ ✓

x • **5** = $5x$ ✓

$9x^2 - 3y$

GCF of $9x^2$ and $3y = 3$

$3 \cdot 3x^2 = 9x^2$ ✓ $\quad 3 \cdot (-y) = -3y$ ✓

$3(3x^2 - y)$

$$3(3x^2-y)$$

Now we have two different variables, an "x" AND "y." Only common variables can be divided from more than one term so the GCF in this situation is the number 3, no variable. The x^2 and y will remain as they are; they have nothing in common, much like most high school couples.

Divide each term by **3**.

$9x^2 \div \mathbf{3} = \mathbf{3x^2}$

$-3y \div \mathbf{3} = \mathbf{-y}$

Answer: **3** (**$3x^2$ - y**)

Check yourself:

3 • **$3x^2$** = $9x^2$ ✓

3 • **-y** = $-3y$ ✓

Factoring with exponents

Trinomials are included below, just TRI it, haha! More on trinomials on page 10.

Factor by GCF PART 1:

1. $4x^2 + 8x$ ______
2. $10x^2 - 15x$ ______
3. $6x + 30y$ ______
4. $5x^2 - 10x$ ______
5. $3x^2 + 12x$ ______
6. $7x^2 - 14y$ ______

TRINOMIAL!

7. $8y^2 - 40y + 4$ ______
8. $5x^2 + 35y^2$ ______
9. $100x^2 - 18x$ ______

TRINOMIAL!

10. $9x^2 + 15x - 3$ ______
11. $5y^2 - 25y$ ______
12. $3x^2 - 21$ ______

TRINOMIAL!

13. $32x^2 + 24x + 16$ ______
14. $15x^3 + 15x^2$ ______
15. $50x^3 - 25x$ ______

Factor by GCF PART 2:

1. $7x^3 + 49x$ ______
2. $8x^{20} - 16x^{10}$ ______
3. $6x + 36xy^3$ ______
4. $15x^6 - 10x^2$ ______
5. $33x^3 + 11x^2$ ______
6. $8x^{14} - 14x^4y$ ______
7. $50x^3 - 25x$ ______
8. $12x^8y^2 + 28xy^2$ ______
9. $10x^5 - 20x^4$ ______
10. $300a^2b^3 + 150a^3b^2$ ______
11. $9x^5y^2 - 12x^2y$ ______
12. $x^8y^2 - 21x^4y^4$ ______
13. $3x^9 + 24x^7 + 18x^3$ ______
14. $2y^6 - 4y^5 + 4y$ ______
15. $7x^3 + 14x^2 - 35x$ ______

FACTORING TRINOMIALS (TRI = 3) $ax^2 + bx + c$ WITH $a = 1$

Trinomials $ax^2 + bx + c$ $a = 1$

How it's done!

Step 1:
Factors of 12 that combine to equal 7:
3 and 4.

Step 2:
In the parentheses,
(x+**3**)(x+**4**).

Step 3:
FOIL and check your work. You'll get the same answer as the initial trinomial
($x^2 + 7x + 12$).

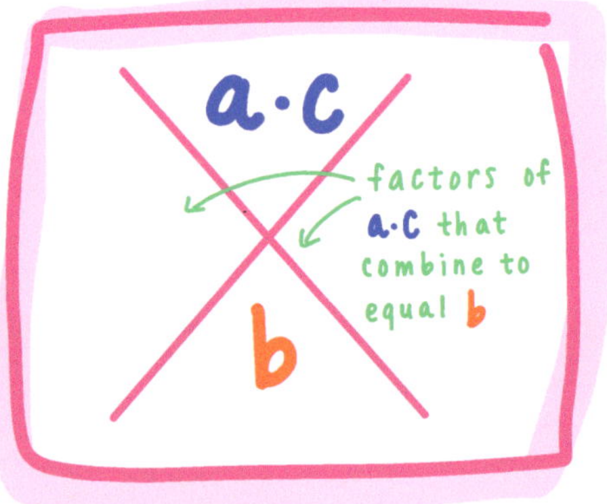

$$x^2 + 7x + 12$$

$a = 1$ $b = 7$ $c = 12$

a·c
$1 \cdot 12$

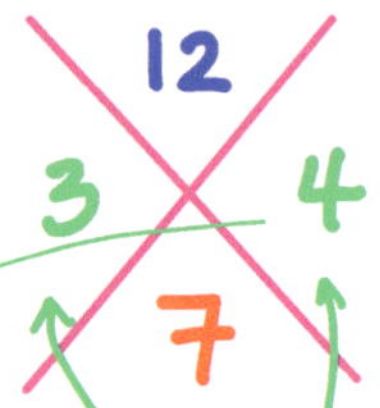

$(x + 3)(x + 4)$

Check
$3 \cdot 4 = 12$ ☑
$3 + 4 = 7$ ☑

factors of 12: $1 \cdot 12$, $2 \cdot 6$, $3 \cdot 4$

CHECK ✓
$(x + 3)(x + 4)$ $x^2 + 4x + 3x + 12 = x^2 + 7x + 12$ ☑

Answer: $(x + 3)(x + 4)$

Trinomials $ax^2 + bx + c$ $a = 1$

Step 1:
Factors of -12 that combine to equal -1:
3 and -4.

Step 2:
In the parentheses,
(x+**3**)(x - **4**).

Step 3:
FOIL and check your work. You'll get the same answer as the initial trinomial
($x^2 - 1x - 12$).

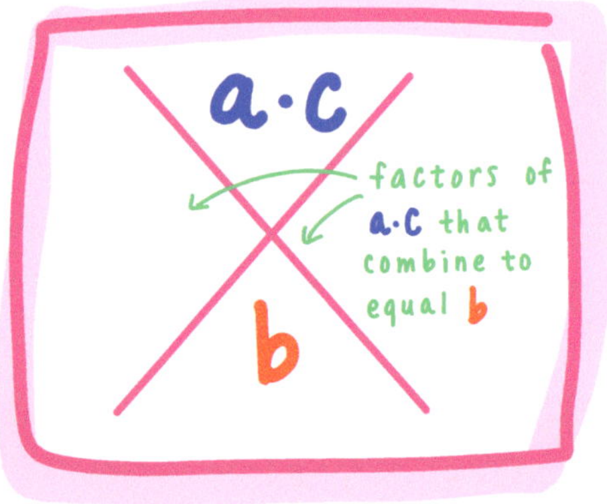

$$x^2 - x - 12$$

$a = 1$ $b = -1$ $c = -12$

a·c
$1 \cdot (-12)$

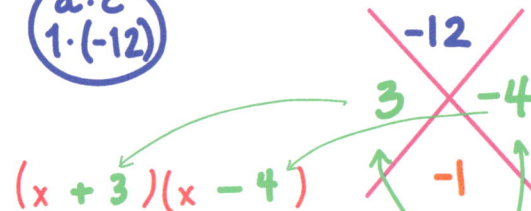

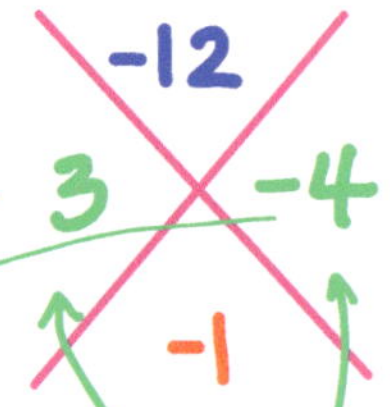

$(x + 3)(x - 4)$

Check
$3 \cdot (-4) = -12$ ☑
$3 + (-4) = -1$ ☑

factors of 12: $1 \cdot 12$, $2 \cdot 6$, $3 \cdot 4$

Check $(x + 3)(x - 4)$ $x^2 - 4x + 3x - 12 = x^2 - x - 12$ ☑

Answer: $(x + 3)(x - 4)$

FOR MORE HELP slide into our DM's @BestMathTutor

Oh, snap... Just kidding, you've got this!

1. $x^2 + 11x + 28$

2. $x^2 + 4x - 12$

3. $x^2 + 13x + 30$

4. $x^2 - 10x - 11$

5. $x^2 - 12x + 35$

6. $x^2 + 10x + 16$

7. $a^2 + 6x + 8$

8. $c^2 - c - 20$

9. $y^2 + 16y + 63$

FACTORING TRINOMALS (TRI = 3) $a^2 + bx + c$ WITH $a > 1$

Trinomials $ax^2 + bx + c$ $a > 1$

Step 1:
Factors of a•c [10] that combine to equal b [7]: ***2 and 5***.

Step 2:
In the parentheses,
x + one factor, and
x + the other factor.
⇨ (x+2)(x+5).

Step 3:
Divide each factor by the "a" value, **2**.
(x + 2/**2**) (x + 5/**2**)

Step 4:
If the fraction does NOT simplify, bring the denominator UP in front of the x: (x+1)(**2**x+5).

Step 5:
FOIL and check your work. You'll get the same answer as the initial trinomial ($2x^2 + 7x + 5$).

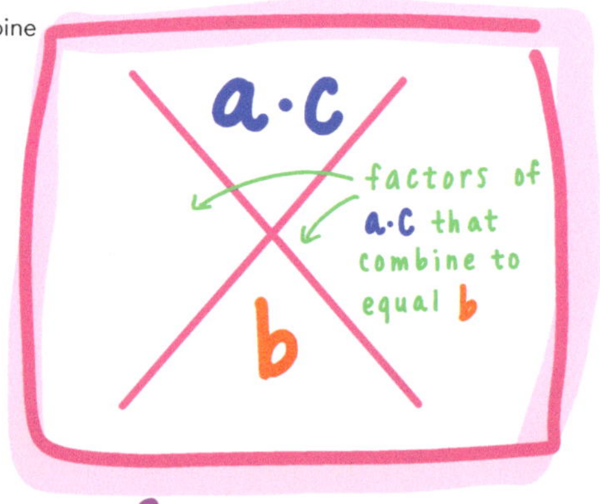

$2x^2 + 7x + 5$

a = 2 b = 7 c = 5

factors of 10: 1·10, 2·5 ✓

a·c
2·5

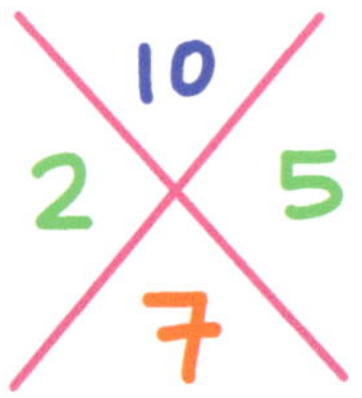

Check
2·5 = 10 ✓
2+5 = 7 ✓

a>1 NEW STEP ÷a on both terms

$(x + 2)(x + 5)$ → $\left(x + \frac{2}{2}\right)\left(x + \frac{5}{2}\right)$

$= \left(x + \frac{2}{2}\right)\left(x + \frac{5}{2}\right) = (x+1)(2x+5)$

✓ $(x+1)(2x+5)$ $2x^2 + 5x + 2x + 5 = 2x^2 + 7x + 5$ ✓

Answer: $(x+1)(2x+5)$

Trinomials $ax^2 + bx + c$ $a > 1$

Step 1:
Factors of -60 that combine to equal -17: ***3 and -20***.

Step 2:
In the parentheses,
x + one factor, and
x + the other factor.
⇨ (x+3)(x - 20).

Step 3:
Divide each factor by the "a" value, **3**.
(x + 3/**3**) (x - 20/**3**)

Step 4:
If the fraction does NOT simplify, bring the denominator UP in front of the x: (x+1)(**3**x - 20).

Step 5:
FOIL and check your work. You'll get the same answer as the initial trinomial ($3x^2 - 17x - 20$).

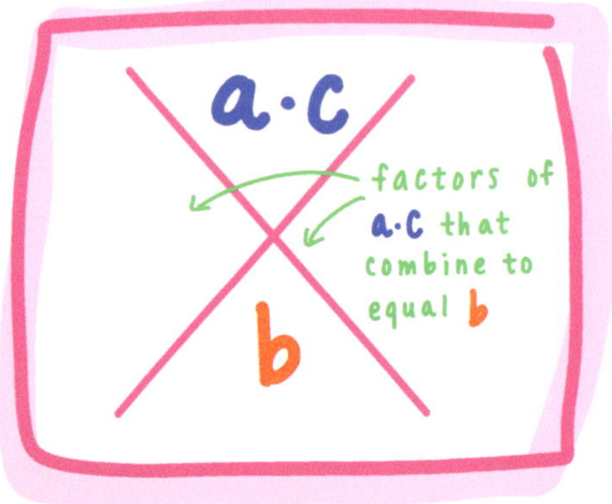

$3x^2 - 17x - 20$

a = 3 b = -17 c = -20

factors of 60:
1,60 2,30 3,20
4,15 5,12 6,10

only combo that multiplies to -60 and adds to -17!

a·c
3·-20

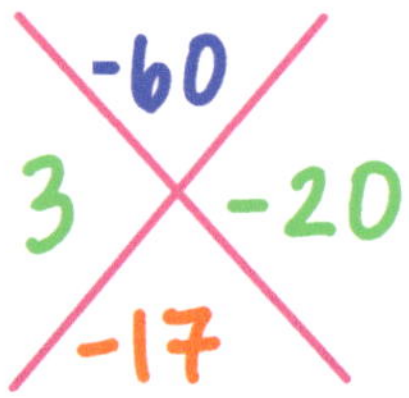

Check
3·(-20) = -60 ✓
3+(-20) = -17 ✓

a>1 NEW STEP ÷a on both terms

$(x + 3)(x - 20)$ → $\left(x + \frac{3}{3}\right)\left(x - \frac{20}{3}\right)$

$= \left(x + \frac{3}{3}\right)\left(x + \frac{-20}{3}\right) = (x+1)(3x-20)$

✓ $(x+1)(3x-20)$ $3x^2 - 20x + 3x - 20 = 3x^2 - 17x - 20$ ✓

Answer: $(x+1)(3x-20)$

If you aren't loving this method of dividing by "a" to complete the factor pair, another method to factoring when a>1 is on the next page!

"Will I use this later in my life?" Probably not... unless you want to write books about it later.

1. $6x^2 - 19x + 15$

2. $4y^2 - 11y + 6$

3. $2a^2 + a - 3$

4. $2x^2 + 3x - 9$

5. $3c^2 - 8c + 4$

6. $5x^2 + 19x + 12$

7. $5a^2 - 18a + 9$

8. $4b^2 - 15b - 25$

9. $6y^2 + 37y + 6$

FACTORING TRINOMALS BY GROUPING (TRI = 3)

The other method when a > 1

Grouping $ax^2 + bx + c$ $a = 1$

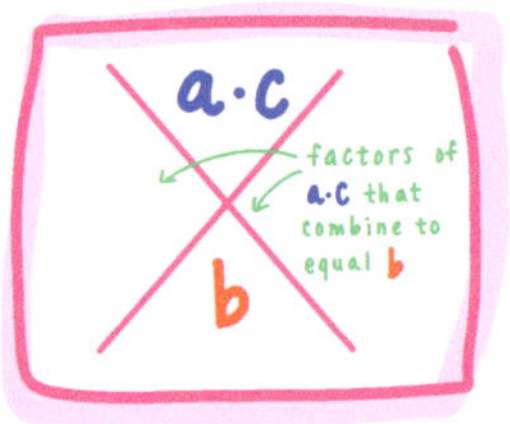

Step 1:
Factors of 12 that combine to equal 7: ***3 and 4***.

Step 2:
Rewrite the polynomial starting with the same "a" value, using the two factors as middle terms, and ending with the "c" value:
$x^2 + \mathbf{3}x + \mathbf{4}x + 12$.

Step 3:
GROUP the first two and the second two terms, find GCF:
$x^2 + \mathbf{3}x$ $+ \mathbf{4}x + 12$
GCF: x   **4**
Divide GCF out of both terms:
x(x + 3)   +**4**(x + 3)

Step 4:
Combine the GCF's as one binomial (**x** + **4**) next to the other binomial (x + 3) to complete factoring (x + 4)(x + 3).

Step 5:
FOIL and check your work. You'll get the same answer as the initial trinomial ($x^2 + 7x + 12$).

$$x^2 + 7x + 12$$

$a = 1$ $b = 7$ $c = 12$

Same as before!

a·c: 1·12

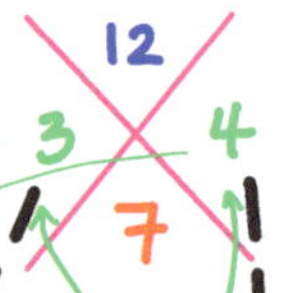

$(x + 3)(x + 4)$

factors of 12: 1·12, 2·6, 3·4

Check: $3 \cdot 4 = 12$ ✓ $3 + 4 = 7$ ✓

$$x^2 + 3x + 4x + 12$$

GCF = x   GCF = 4

factor in groups of 2 by GCF

$$x(x + 3) + 4(x + 3)$$

$$(x + 4)(x + 3)$$

These MUST BE exactly the SAME! x+3 x+3 ✓

Answer: $(x + 4)(x + 3)$

(same problem as page 10)

Grouping $ax^2 + bx + c$ $a > 1$

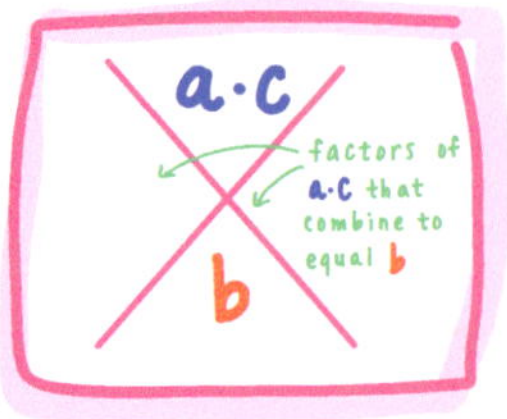

Step 1:
Factors of -60 that combine to equal -17: ***3 and -20***.

Step 2:
Rewrite the polynomial starting with the same "a" value, using the two factors as middle terms, and ending with the "c" value:
$3x^2 + \mathbf{3}x - \mathbf{20}x + 12$.

Step 3:
GROUP the first two and the second two terms, find GCF:
$3x^2 + \mathbf{3}x$ $- \mathbf{20}x - 20$
GCF: x   **20**
Divide GCF out of both terms:
3x(x + 1)   - **20**(x + 1)

Step 4:
Combine the GCF's as one binomial (**3x** - **20**) next to the other binomial (x + 1) to complete factoring (3x - 20)(x + 1).

Step 5:
FOIL and check your work. You'll get the same answer as the initial trinomial ($3x^2 - 17x - 20$).

$$3x^2 - 17x - 20$$

$a = 3$ $b = -17$ $c = -20$

Same as before!

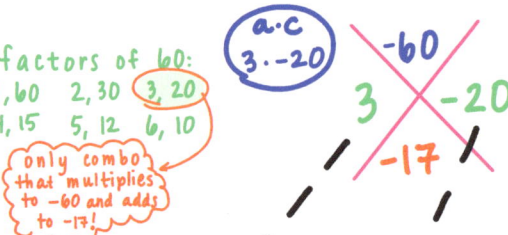

factors of 60:
1, 60   2, 30   3, 20
4, 15   5, 12   6, 10

only combo that multiplies to -60 and adds to -17!

a·c: 3·-20

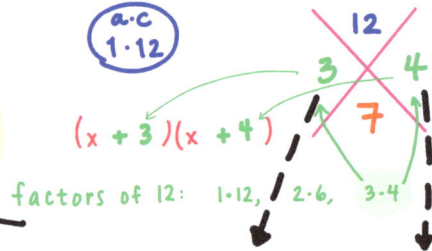

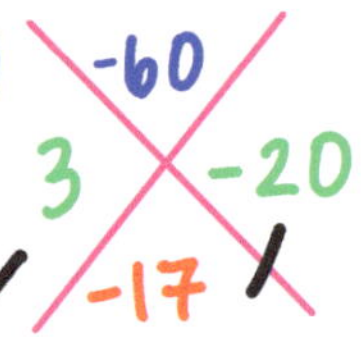

Check: $3 \cdot (-20) = -60$ ✓ $3 + (-20) = -17$ ✓

$$3x^2 + 3x - 20x - 20$$

GCF = 3x   GCF = -20

factor in groups of 2 by GCF

$$3x(x + 1) - 20(x + 1)$$

$$(3x - 20)(x + 1)$$

x+1 x+1 ✓

Answer: $(3x - 20)(x + 1)$

(same problem as page 12)

Same as p. 15, but factor by grouping this time:

1. $6x^2 - 19x + 15$

2. $4y^2 - 11y + 6$

3. $2a^2 + a - 3$

4. $2x^2 + 3x - 9$

5. $3c^2 - 8c + 4$

6. $5x^2 + 19x + 12$

7. $5a^2 - 18a + 9$

8. $4b^2 - 15b - 25$

9. $6y^2 + 37y + 6$

SPECIAL CASE: SUM $a^2 + b^2$ AND DIFFERENCE $a^2 - b^2$ OF SQUARES² (BINOMIALS)

Difference of Squares

These are SUPER EASY! Simply square root the "a" and "b" term and write the result as a set of +/- factors:

$x^2 - y^2$ ⇨⇨⇨ $\sqrt{x^2} = \mathbf{x}$ and $\sqrt{y^2} = \mathbf{y}$. Now put the result [**x** and **y**] in a set of parentheses with one being addition and one being subtraction.
(x + y)(x - y).

$a^2 - 4$ ⇨⇨⇨ $\sqrt{a^2} = \mathbf{a}$ and $\sqrt{4} = \mathbf{2}$. Now put the result [**a** and **2**] in a set of parentheses with one being addition and one being subtraction.
(a + 2)(a - 2).

$c^2 - 25$ ⇨⇨⇨ $\sqrt{c^2} = \mathbf{c}$ and $\sqrt{25} = \mathbf{5}$. Now put the result [**c** and **5**] in a set of parentheses with one being addition and one being subtraction.
(c + 5)(c - 5).

The reason there is no middle term "bx" in difference of squares is because when you FOIL the factors to check your answer, you will see that the middle term is eliminated!

(x + y)(x - y) ⇨ $x^2 + xy - xy - y^2$ ⇨⇨⇨ x^2 **+ xy - xy** $- y^2$ ⇨⇨⇨ $x^2 - y^2$

(a + 2)(a - 2) ⇨ $a^2 + 2a - 2a - 4$ ⇨⇨⇨ a^2 **+ 2a - 2a** $- 4$ ⇨⇨⇨ $a^2 - 4$

(c + 5)(c - 5) ⇨ $c^2 + 5c - 5c - 25$ ⇨⇨⇨ c^2 **+ 5c - 5c** $- 25$ ⇨⇨⇨ $c^2 - 25$

Difference of Squares $a^2 - b^2$

$$a^2 - b^2 = (a + b)(a - b)$$

$x^2 - y^2 = (x + y)(x - y)$

$a^2 - 4 = (a + 2)(a - 2)$

$c^2 - 25 = (c + 5)(c - 5)$

$36 - x^2 = (6 + x)(6 - x)$

$49y^2 - 16 = (7y + 4)(7y - 4)$

$49y^4 - 16x^2 = (7y^2 + 4x)(7y^2 - 4x)$

$(a+b)^2 - 25$

$= [(a+b) + 5][(a+b) - 5]$

$= (a + b + 5)(a + b - 5)$

$(3x + y)^2 - 36$

$= [(3x+y) + 6][(3x + y - 6]$

$= (3x + y + 6)(3x + y - 6)$

Sum of Squares

Always **PRIME**, meaning they cannot be factored. In these types of problems, simply write "prime" as your answer. The problems **IN BLUE** to the right are examples of sums of squares, all of which cannot be factored with REAL numbers.

The reason these "do not work" is because if you set them equal to zero (as you must do to "solve" or "find zeros" or "find x-intercepts" or "find roots"), you will end up with **(term)² = a NEGATIVE number**.

It is not possible to take the square root of a negative number and get a REAL answer. *It is possible to get imaginary numbers, but that is a different topic.*

For example:

$c^2 + 25 = 0$	Subtract 25 from both sides
$c^2 = -25$	Square root both sides to solve for c
$c = \sqrt{-25}$	It is not possible to get a real solution

Sum of Squares $a^2 + b^2$

$x^2 + y^2$ $a^2 + 4$ $c^2 + 25$ $4x^2 + 36$

$49y^2 + 16$ $9b^2 + c^2$ $x^2 + 64$ $16x^2 + 9y^2$

sums of squares are **ALWAYS PRIME**

do NOT factor

FOR MORE HELP slide into our DM's @BestMathTutor

Go!

Factor Squares PART 1:

1. $4x^2 - 9$ ________________
2. $100x^2 - 81$ ________________
3. $25x^2 + 9$ ________________
4. $a^2 - 121$ ________________
5. $9y^2 - 4z^2$ ________________
6. $144 - x^2$ ________________
7. $49a^2 - 36b^2$ ________________
8. $64b^2 + 9$ ________________
9. $64b^2 - 9$ ________________

GCF FIRST!

10. $8x^2 - 50$ ________________

GCF FIRST!

11. $45y^2 - 125$ ________________

GCF FIRST!

12. $3x^2 - 27$ ________________
13. $(x + y)^2 - 9$ ________________
14. $(7a - 4b)^2 - 25$ ________________
15. $(11x + 6y)^2 - 16$ ________________

Factor Squares PART 2:

1. $16x^2 - 1$ ________________
2. $121x^2 - 64$ ________________
3. $4a^2 - 9b^2$ ________________
4. $49a^2 - 121$ ________________
5. $144y^2 - z^2$ ________________
6. $400 - x^2$ ________________
7. $36x^2 + 9$ ________________
8. $625b^2 + 9$ ________________
9. $625b^2 - 9$ ________________
10. $100x^2 - 81$ ________________
11. $49y^2 - 121$ ________________
12. $x^2 - 144$ ________________
13. $(5x + 2y)^2 - 16$ ________________
14. $(a - 4b)^2 - 9$ ________________
15. $(3x + 4y)^2 - 49$ ________________

SPECIAL CASE: SUM AND DIFFERENCE OF CUBES³ (BINOMIALS) $a^3 + b^3$, $a^3 - b^3$

Notice the similarities btw sum and difference of cubes. Remember the template and SOAP and you're good to go!

Difference of Cubes

These are SUPER EASY! Simply cube root the "a" and "b" term and use those in the following template: $a^3 - b^3 =$ **$(a - b)(a^2 + ab + b^2)$**

The tool to remember the sign on each term is **SOAP**

Same - **O**pposite - **A**lways **P**ositive

$x^3 - y^3$ ⇨⇨⇨ $\sqrt[3]{x^3} =$ **x** and $\sqrt[3]{y^3} =$ **y**. Now, a = **x** and b = **y** into the template for cubes: $(x - y)(x^2 + xy + y^2)$. DONE.

(same, opposite, always positive)

$8x^3 - 27y^3$ ⇨⇨⇨ $\sqrt[3]{8x^3} =$ **$2x$** and $\sqrt[3]{27y^3} =$ **$3y$**. Now, a = **2x** and b = **3y** into the template for cubes: $(2x - 3y)(4x^2 + 6xy + 9y^2)$. DONE.

(same, opposite, always positive)

The resulting trinomial DOES NOT FACTOR any further.
The binomial and trinomial factors are your **final answer**.

Sum of Cubes

The same as difference of cubes but following the **SOAP** rule. Simply cube root the "a" term and "b" term and use those in the following template: $a^3 + b^3 =$ **$(a + b)(a^2 - ab + b^2)$**

$x^3 + y^3$ ⇨⇨⇨ $\sqrt[3]{x^3} =$ **x** and $\sqrt[3]{y^3} =$ **y**. Now put the result [**x** and **y**] in the template for cubes: $(x + y)(x^2 - xy + y^2)$. DONE.

(same, opposite, always positive)

$8x^3 + 27y^3$ ⇨⇨⇨ $\sqrt[3]{8x^3} =$ **$2x$** and $\sqrt[3]{27y^3} =$ **$3y$**. Now put the result [**2x** and **3y**] in the template for cubes: $(2x + 3y)(4x^2 - 6xy + 9y^2)$. DONE.

(same, opposite, always positive)

The resulting trinomial DOES NOT FACTOR any further.
The binomial and trinomial factors are your **final answer**.

Difference of Cubes $a^3 - b^3$

SOAP SOAP SOAP

$$a^3 - b^3 = (a - b)(a^2 + ab + b^2)$$

the sign is: Same Opposite Always Positive — S (same), O (opposite), AP (always positive)

ANSWER

$$x^3 - y^3 = (x - y)(x^2 + xy + y^2)$$

$$8x^3 - 27y^3 = (2x - 3y)(4x^2 + 6xy + 9y^2)$$

$$a^3 - 64 = (a - 4)(a^2 + 4a + 16)$$

$$16x^3 - 250 \quad GCF = 2 \quad 2(8x^3 - 125) = 2(2x - 5)(4x^2 + 10x + 25)$$

resulting trinomial DOES NOT FACTOR any further

Sum of Cubes $a^3 + b^3$

SOAP SOAP SOAP

$$a^3 + b^3 = (a + b)(a^2 - ab + b^2)$$

the sign is: Same Opposite Always Positive — S (same), O (opposite), AP (always positive)

ANSWER

$$x^3 + y^3 = (x + y)(x^2 - xy + y^2)$$

$$8x^3 + 27y^3 = (2x + 3y)(4x^2 - 6xy + 9y^2)$$

$$a^3 + 64 = (a + 4)(a^2 - 4a + 16)$$

$$16x^3 + 250 \quad GCF = 2 \quad 2(8x^3 + 125) = 2(2x + 5)(4x^2 - 10x + 25)$$

resulting trinomial DOES NOT FACTOR any further

+/– CUBES^{333} It's not as bad as it looks!

Factor:

1. $x^3 - 27$ ____________________
2. $1000y^3 - 8$ ____________________
3. $125a^3 + 343$ ____________________
4. $c^3 - 64$ ____________________
5. $8y^3 + z^3$ ____________________
6. $x^3 - 512$ ____________________
7. $343a^3 - 27b^3$ ____________________
8. $p^3 + 125$ ____________________
9. $64x^3 + 729$ ____________________
10. $27x^3 - 125$ ____________________
11. $512y^3 + 125$ ____________________
12. $343x^3 - 27y^3$ ____________________
13. $a^3 - 64b^3$ ____________________
14. $729x^3 + 125y^3$ ____________________
15. $250t^3 - 16r^3$ ____________________

SPECIAL CASE: $a^2 + 2ab + c^2 = (a + b)^2$ $a^2 - 2ab + c^2 = (a - b)^2$

Recognizing a trinomial that factors to a binomial squared

The **1st** and **3rd** term will be **SQUARES**^{222}:

Ex: $x^2, y^2, a^2, b^2, 9x^2, 49y^2, 4a^2$, 16, 25, 36, 49, 64, 81, 100

The **3rd term** will always be positive + + + + + + + + + + + +
Keep these two things in mind and you will be able to recognize a trinomial that factors to a binomial squared.

$x^2 + 14x + 49$ When you see a 14 and 49, start thinking 7

$\mathbf{7} \bullet 2 = 14$ and $\mathbf{7^2} = 49$

Once you recognize that the trinomial factors to a binomial squared, take the square root of the first and third term to get "a" and "b" term $(\mathbf{a} +/- \mathbf{b})^2$.

$\sqrt{x^2} = \mathbf{x}$ and $\sqrt{49} = \mathbf{7}$. $a = \mathbf{x}$, $b = \mathbf{7}$.

The sign on the b term is the same sign as the middle term.

In this example, it is positive, so the factored answer is $\mathbf{(x + 7)^2}$.

$x^2 - 16x + 64$ When you see a 16 and 64, start thinking 8

$\mathbf{8} \bullet 2 = 16$ and $\mathbf{8^2} = 64$

Once you recognize that the trinomial factors to a binomial squared, take the square root of the first and third term to get "a" and "b" term $(\mathbf{a} +/- \mathbf{b})^2$.

$\sqrt{x^2} = \mathbf{x}$ and $\sqrt{64} = \mathbf{8}$. $a = \mathbf{x}$, $b = \mathbf{8}$.

The sign on the b term is the same sign as the middle term.

In this example, it is negative, so the factored answer is $\mathbf{(x - 8)^2}$.

$9x^2 + 12xy + 4y^2$ When you see a 9, start thinking 3
When you see a 4, start thinking 2.

Think of the possible square roots and be familiar with common squares of 2^2 - 10^2.

Once you recognize that the trinomial factors to a binomial squared, take the square root of the **1st** and **3rd** term to get "a" and "b" terms $(\mathbf{a} +/- \mathbf{b})^2$.

$\sqrt{9x^2} = \mathbf{3x}$ and $\sqrt{4y^2} = \mathbf{2y}$. $a = \mathbf{3x}$, $b = \mathbf{2y}$.

$2ab = 2(3x)(2y) = 12xy$ LOOK! THAT'S THE MIDDLE TERM!

The sign on the b term is the same sign as the middle term.

In this example, it is positive, so the factored answer is $\mathbf{(3x + 2y)^2}$.

Binomial Squared

$a^2 + 2ab + b^2 = (a + b)^2$

Look for squares in 1st and 3rd term

ex: 4, 9, 16, 25, 36, 49, etc.

The 3rd term is always positive! +

$x^2 + 14x + 49 = (x + 7)^2$

$x^2 - 14x + 49 = (x - 7)^2$

$9x^2 - 12xy + 4y^2 = (3x - 2y)^2$

$9x^2 + 12xy + 4y^2 = (3x + 2y)^2$

$x^2 + 16x + 64 = (x + 8)^2$

$x^2 - 16x + 64 = (x - 8)^2$

$25a^2 - 60ab + 36b^2 = (5a - 6b)^2$

$25a^2 + 60ab + 36b^2 = (5a + 6b)^2$

$x^2 + 8x + 16 = (x + 4)^2$

$x^2 - 8x + 16 = (x - 4)^2$

$4x^2 - 36xy + 81y^2 = (2x - 9y)^2$

$4x^2 + 36xy + 81y^2 = (2x + 9y)^2$

Factoring special case – binomial squared^2

Factor:

1. $x^2 + 10x + 25$ ______
2. $x^2 - 10x + 25$ ______
3. $y^2 + 12x + 36$ ______
4. $y^2 - 12y + 36$ ______
5. $16a^2 - 40a + 25$ ______
6. $16a^2 + 40a + 25$ ______
7. $x^2 - 6xy + 9y^2$ ______
8. $x^2 + 6xy + 9y^2$ ______
9. $9x^2 - 48xy + 64y^2$ ______
10. $9x^2 + 48xy + 64y^2$ ______
11. $4a^2 + 8ab + 4b^2$ ______
12. $x^2 - 2xy + y^2$ ______
13. $25x^2 + 70xy + 49y^2$ ______

HINT: GCF FIRST!

14. $2x^2 + 20x + 50$ ______
15. $8x^2 + 16xy + 8y^2$ ______

PRIME: CANNOT BE FACTORED

PRIME = does not factor

$$x^2 + 2x + 7$$

$a \cdot c = 1 \cdot 7 = 7$ no factors of 7 combine to equal 2

$$x^2 + 7x + 11$$

$a \cdot c = 1 \cdot 11 = 11$ no factors of 11 combine to equal 7

$$2x^2 - 5x + 8$$

$a \cdot c = 2 \cdot 8 = 16$ no factors of 16 combine to equal -5

$$5x^2 + 3x - 11$$

$a \cdot c = 5 \cdot (-11) = -55$ no factors of -55 combine to equal 3

$16x^2 + 49y^2$

$a^2 + b^2$

Sum of Squares **ALWAYS PRIME**

If you have a problem that cannot be factored, it is PRIME.

Be very careful and do not be quick to assume a polynomial is PRIME.

Double check your factor pairs to make sure you didn't miss one!

SUMMARY REVIEW

Factors are terms that, when multiplied, equal the same term.

GRAPHING CALCULATOR TIP: To easily find all of the factors of a number, go to **y=** in your graphing calculator and **enter the number/x**.
For example: y = 36/x y = 343/x y = -80/x y = 125/x
Then go to **TABLE** and *view all of the factor pairs!*

When factoring by GCF, divide the GCF from all terms:

6x + 2	GCF = 2,	2(3x+1)	*redistribute to check answer*
$8x^2 + 6x$	GCF = 2x,	2x(4x + 3)	*redistribute to check answer*

Factoring trinomials:

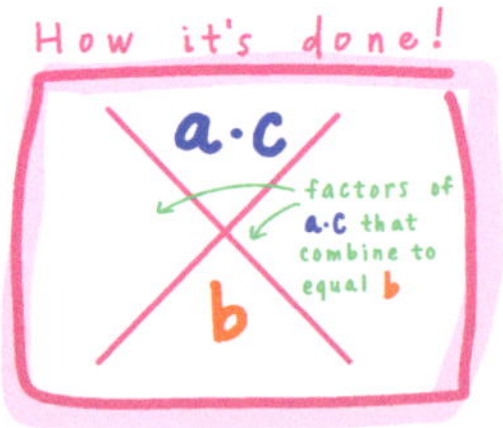

Step 1: a•c and find factors of that amount that combine to equal b.
Step 2: If a = 1, simply write the factors in parentheses with x: (x + 2)(x + 3).
If a >1, factor by GROUPING (p.16)
or divide each factor by a (p.14).
Always redistribute or FOIL the factors to check that the answer is correct.

Sum and Difference of SQUARES²

$a^2 + b^2$

SUMS OF SQUARES are ALWAYS PRIME!!! They cannot be factored.

$a^2 - b^2$

Step 1: take the square root both terms.
Step 2: Put the result in a set of parentheses PLUS/MINUS **(a + b)(a - b)**

Sum and Difference of CUBES³ (SOAP)

$a^3 + b^3$

Step 1: Find the cube root of the first and second term to give you a and b.
Step 2: Plug those values into the template: $(a + b)(a^2 - ab + b^2)$

$a^3 - b^3$

Step 1: Find the cube root of the first and second term to give you a and b.
Step 2: Plug those values into the template: $(a - b)(a^2 + ab + b^2)$

Special Case - BINOMIAL SQUARED $(a + b)^2$ and $(a - b)^2$

Become familiar with the squares of 2^2 to 10^2 and be aware of them.
You will start to notice patterns and recognize squared binomials easily.
The **1st** and **3rd** term are always **SQUARES²** $a^2 + 2ab + b^2$ or $a^2 - 2ab + b^2$.
The **3rd term** will always be positive **+++**

You're at mile 25... you can do this. Full Review

Find the GCF of the following:

1. *a.* 15 and 25 *b.* 8 and 34
2. *a.* $12x^2$ and $8x$ *b.* y^2 and $6z$

Factor by GCF:

3. *a.* $8x - 18$ *b.* $25a^2 + 10a$

   c. $5b^2 + 35b$ *d.* $45x^3 + 9x^2 + 9x$

Factor:

4. $a^2 - a - 20$
5. $x^2 + 6x + 8$
6. $y^2 + 2y + 9$
7. $n^2 - 10n + 25$
8. $x^2 - 7x - 30$
9. $2a^2 - 4a - 70$
10. $y^2 - 11y + 28$
11. $x^2 + 14x + 40$
12. $x^2 + 14x + 45$
13. $z^2 - z - 56$
14. $a^2 - 81$
15. $5x^2 - 18x + 9$
16. $8a^3 - 27$
17. $3n^2 + 8n - 5$
18. $64x^2 - 49y^2$
19. $3a^2 - 17a - 20$
20. $7y^2 + 50y + 7$
21. $36x^2 + 9$
22. $(5x + 6y)^2 - 49$
23. $81a^2 - 100b^2$
24. $2x^3 + 16y^3$
25. $9y^2 - 30yz + 25z^2$
26. $8x^2 - 12x - 8$
27. $a^2 + 2ab + b^2$
28. $27z^3 - 343$
29. $5c^2 - 11x - 12$
30. $25x^2 + 30x + 9$

EXTRA PRACTICE... BC YOU'RE THAT AWESOME

Factor:

1. $5x^3 + 40y^3$
2. $16a^2 + 56a + 49$
3. $2x^2 - 16x + 32$
4. $8y^2 - 16 - 28y$
5. $25 - 10b + b^2$
6. $6a^6 + a^3 - 2$
7. $81 - z^4$
8. $(x + y)^2 - 25$
9. $y^2 - 144$
10. $25 - a^2$
11. $5x^2y^3 - 15x^3y^2$
12. $x^2 + 9x + 20$
13. $c^3 + 9c^2$
14. $y^3 + 8$
15. $64x^4 + x$
16. $r^6 - 64$
17. $6y^2 + 23y + 20$
18. $8r^2 - 6r - 9$
19. $3x + x^2 - 10$
20. $a^2 + 5a - 84$
21. $36 - (x - y)^2$
22. $9a^3 - 49a$
23. $6x^2 - 7x - 10$
24. $(1/4) - x^2$
25. $64x^3 + 27$
26. $81 - 18a + a^2$
27. $y^2 - 12y + 36$
28. $121z^2 - 1$
29. $6c^2 + 12c + 6$
30. $25x^2 + 36y^2$
31. $t^2 - 8t - 48$
32. $11a^2 - 11b^2$
33. $3q^5 - 12q^3$
34. $8xy^4 - 8x^4y$
35. $5x^2 - 2x + 3$
36. $y^3 - 343$
37. $36x^2 - 9$
38. $a^2 + 5a - 36$
39. $216 - x^3$
40. $9c^5 + 99c^3d^5$
41. $f^2 - 5f - 14$
42. $9x^2y^2 - 25y^4$
43. $s^2 - 3s - 2$
44. $6y^3 + 48$
45. $3x^2 - 34x - 24$
46. $x^2 + 8x + 16$
47. $x^6 - 1$
48. $27a^2 - 30a - 8$

WORKSPACE | NOTES | REMINDERS | REVIEW POINTS

ANSWERS

Find the factors of (PAGE 3)

1.	30:	1•30, 2•15, 3•10, 5•6	8.	100:	1•100, 2•50, 4•25, 5•20, 10•10
2.	45:	1•45, 3•15, 5•9	9.	-8:	-1•8, 1•-8, -2•4, 2•-4
3.	6:	1•6, 2•3	10.	-20:	-1•20, 1•-20, -2•10, 2•-10, -4•5, 4•-5
4.	8:	1•8, 2•4	11.	-12:	-1•12, 1•-12, -2•6, 2•-6, -3•4, 3•-4
5.	19:	1•19	12.	-50:	-1•50, 1•-50, -2•25, 2•-25, -5•10, 5•-10
6.	25:	1•25, 5•5	13.	-80:	-1•80, 1•-80, -2•40, 2•-40, -4•20, 4•-20, -5•16, 5•-16, -8•10, 8•-10
7.	64:	1•64, 2•32, 4•16, 8•8	14.	-6:	-1•6, 1•-6, -2•3, 2•-3
			15.	-25:	-1•25, 1•-25, -5•5

Remember:
The GCF is simply the largest term that can be divided out of two or more terms.

If it helps, list all of the factors of each term (as I did here to the left) and circle or highlight the largest factor the terms have in common.

Once you get more comfortable with GCF, you will recall factors quickly and mentally come up with the GCF in no time!

You're doing great!

Find the GCF (PAGE 5, PART 1)

1. **4 and 12** — 4: 1•**4**, 2•2 — 12: 1•12, 2•6, 3•**4** — GCF = **4**
2. **9 and 3** — 9: 1•9, 3•**3** — 3: 1•**3** — GCF = **3**
3. **8 and 20** — 8: 1•8, 2•**4** — 20: 1•20, 2•10, **4**•5 — GCF = **4**
4. **16 and 30** — 16: 1•16, **2**•8, 4•4 — 30: 1•30, **2**•15, 3•10 — GCF = **2**
5. **14, 21 and 28** — 14: 1•14, 2•**7** — 21: 1•21, 3•**7** — 28: 1•28, 2•14, 4•**7** — GCF = **7**
6. **18, 3 and 27** — 18: 1•18, 2•9, **3**•6 — 3: 1•**3** — 27: 1•27, **3**•9 — GCF = **3**
7. **64, 24, and 8** — 64: 1•64, 2•32, 4•16, **8**•8 — 24: 1•24, 2•12, 3•**8**, 4•6 — 8: 1•**8**, 2•4 — GCF= **8**
8. **2x and 10** — 2x: 1x•**2**, 2x•1 — 10: 1•10, **2**•5 — GCF = **2**
9. **4x and 8** — 4x: 1x•4, 4x•1, 2x•**2**, 2•2x — 8: 1•8, **2**•2 — GCF = **2**
10. **3x and 12** — 3x: 1x•**3**, 3x•1 — 12: 1•12, 2•6, **3**•4 — GCF = **3**
11. **14x and 21** — 14x: 1x•14, 1•14x, 2x•**7**, 2•7x — 21: 1•21, 3•**7** — GCF = **7**
12. **20x, 30 and 90** — 20x: 1x•20, 1•20x, 2x•**10**, 2•10x, 4x•5, 4•5x — 30: 1•30, 2•15, 3•**10**, 5•6 — 90: 1•90, 2•45, 3•30, 5•18, 6•15, 9•**10** — GCF = **10**
13. **15x, 5 and 40** — 15x: 1x•15, 1•15x, 3x•**5**, 3•5x — 5: 1•**5** — 40: 1•40, 2•20, 4•10, **5**•8 — GCF = **5**
14. **8x, 16 and 32** — 8x: 1x•**8**, 1•8x, 2x•4, 2•4x — 16: 1•16, 2•**8**, 4•4 — 32: 1•32, 2•16, 4•**8** — GCF = **8**
15. **11x, 22 and 99** — 11x: 1x•**11** , 1•11x — 22: 1•22, 2•**11** — 99: 1•99, 3•33, 9•**11** — GCF = **11**

Part 2 →

REMEMBER:

Calculator tip

Go to Y=

example: y = 45/x

go to TABLE

& view all factors of 45!

Factor by GCF (PAGE 7, PART 1)

1.	$4x + 2$	GCF = 2	$4x \div 2 = 2x$	$2 \div 2 = 1$	$2(2x + 1)$
2.	$8x - 2$	GCF = 2	$8x \div 2 = 4x$	$-2 \div 2 = -1$	$2(4x - 1)$
3.	$9x + 30$	GCF = 3	$9x \div 3 = 3x$	$30 \div 3 = 10$	$3(3x + 10)$
4.	$15x - 10$	GCF = 5	$15x \div 5 = 3x$	$-10 \div 5 = -2$	$5(3x - 2)$
5.	$30x + 25$	GCF = 5	$30x \div 5 = 6x$	$25 \div 5 = 5$	$5(6x + 5)$
6.	$2x - 12$	GCF = 2	$2x \div 2 = 1x$	$-12 \div 2 = -6$	$2(x - 6)$
7.	$21x - 14$	GCF = 7	$21x \div 7 = 3x$	$-14 \div 7 = -2$	$7(3x - 2)$
8.	$7x + 35$	GCF = 7	$7x \div 7 = 1x$	$35 \div 7 = 5$	$7(x + 5)$
9.	$100x - 45$	GCF = 5	$100x \div 5 = 20x$	$-45 \div 5 = -9$	$5(20x - 9)$
10.	$45x + 9$	GCF = 9	$45x \div 9 = 5x$	$9 \div 9 = 1$	$9(5x + 1)$
11.	$5x - 20$	GCF = 5	$5x \div 5 = 1x$	$-20 \div 5 = -4$	$5(x - 4)$
12.	$6x - 22$	GCF = 2	$6x \div 2 = 3x$	$-22 \div 2 = -11$	$2(3x - 11)$
13.	$27x + 18$	GCF = 9	$27x \div 9 = 3x$	$18 \div 9 = 2$	$9(3x + 2)$
14.	$8x + 14$	GCF = 2	$8x \div 2 = 4x$	$14 \div 2 = 7$	$2(4x + 7)$
15.	$11x - 33$	GCF = 11	$11x \div 11 = 1x$	$-33 \div 11 = -3$	$11(x - 3)$

Find the GCF (PAGE 5, PART 2):

1.	x^2 and xy	GCF = x
2.	$6a^3$ and $3a^2$	GCF = $3a^2$
3.	$3x$ and $9y$	GCF = 3
4.	$20c^3$ and $15cd^3$	GCF = $5c$
5.	$2x^{12}$ and $2x^{11}$	GCF = $2x^{11}$
6.	$8a^5$ and $17a^3b^2$	GCF = a^3
7.	$14d^2e^4f^2$ and $7d^2e^3f$	GCF = $7d^2e^3f$
8.	$5x^{20}$ and $10x^{19}$	GCF = $5x^{19}$
9.	$6x^3$, $3x^2$ and $9x$	GCF = $3x$
10.	$3y^5$, $18y^4$ and $12y^8$	GCF = $3y^4$
11.	$25a^3$,$15ab$ and $18a^2$	GCF = a
12.	x^4, $4x^5$ and $8x^3$	GCF = x^3
13.	$15x^3$, $5x^2$ and $20x$	GCF = $5x$
14.	$8ab^4c^3$, $16a^3bc^3$ and $32a^2bc^3$	GCF = $8abc^3$
15.	$11x^3$, $22x^2$ and $99x$	GCF = $11x$

Factor by GCF (PAGE 7, PART 2)

1.	$9x + 3$	GCF = 3	$9x \div 3 = 3x$	$3 \div 3 = 1$	$3(3x + 1)$
2.	$11x - 121$	GCF = 11	$11x \div 11 = 1x$	$-121 \div 11 = -11$	$11(x - 11)$
3.	$6x + 30$	GCF = 6	$6x \div 6 = 1x$	$30 \div 6 = 5$	$6(x + 5)$
4.	$12x - 10$	GCF = 2	$12x \div 2 = 6x$	$-10 \div 2 = -5$	$2(6x - 5)$
5.	$36x + 12$	GCF = 12	$36x \div 12 = 3x$	$12 \div 12 = 1$	$12(3x + 1)$
6.	$7x - 14$	GCF = 7	$7x \div 7 = 1x$	$-14 \div 7 = -2$	$7(x - 2)$
7.	$28x - 40$	GCF = 4	$28x \div 4 = 7x$	$-40 \div 4 = -10$	$4(7x - 10)$
8.	$5x + 35$	GCF = 5	$5x \div 5 = 1x$	$35 \div 5 = 7$	$5(x + 7)$
9.	$100x - 18$	GCF = 2	$100x \div 2 = 50x$	$-18 \div 2 = -9$	$2(50x - 9)$
10.	$40x + 15$	GCF = 5	$45x \div 5 = 8x$	$15 \div 5 = 3$	$5(8x + 3)$
11.	$5x - 25$	GCF = 5	$5x \div 5 = 1x$	$-25 \div 5 = -5$	$5(x - 5)$
12.	$3x - 21$	GCF = 3	$3x \div 3 = 1x$	$-21 \div 3 = -7$	$3(x - 7)$
13.	$32x + 24$	GCF = 8	$32x \div 8 = 4x$	$24 \div 8 = 3$	$8(4x + 3)$
14.	$15x + 15$	GCF = 15	$15x \div 15 = 1x$	$15 \div 15 = 1$	$15(x + 1)$
15.	$50x - 25$	GCF = 25	$50x \div 25 = 2x$	$-25 \div 25 = -1$	$25(2x - 1)$

ANSWERS

Factor by GCF (PAGE 9, PART 1)

1.	$4x^2 + 8x$	GCF = $4x$	$4x^2 \div 4x = 1x$	$8x \div 4x = 2$	$4x(x + 2)$
2.	$10x^2 - 15x$	GCF = $5x$	$10x^2 \div 5x = 2x$	$-15x \div 5x = -3$	$5x(2x - 3)$
3.	$6x + 30y$	GCF = 6	$6x \div 6 = 1x$	$30y \div 6 = 5y$	$6(x + 5y)$
4.	$5x^2 - 10x$	GCF = $5x$	$5x^2 \div 5x = 1x$	$-10x \div 5x = -2$	$5x(x - 2)$
5.	$3x^2 + 12x$	GCF = $3x$	$3x^2 \div 3x = 1x$	$12x \div 3x = 4$	$3x(x + 4)$
6.	$7x^2 - 14y$	GCF = 7	$7x^2 \div 7 = 1x^2$	$-14y \div 7 = -2y$	$7(x^2 - 2y)$
7.	$8y^2 - 40y + 4$	GCF = 4	$4(2y^2 - 10y + 1)$		
8.	$5x^2 + 35y^2$	GCF = 5	$5x^2 \div 5 = 1x^2$	$35y^2 \div 5 = 7y^2$	$5(x^2 + 7y^2)$
9.	$100x^2 - 18x$	GCF = $2x$	$100x^2 \div 2x = 50x$	$-18x \div 2x = -9$	$2x(50x - 9)$
10.	$9x^2 + 15x - 3$	GCF = 3	$3(3x^2 + 5x - 1)$		
11.	$5y^2 - 25y$	GCF = $5y$	$5y^2 \div 5y = 1y$	$-25y \div 5y = -5$	$5y(y - 5)$
12.	$3x^2 - 21$	GCF = 3	$3x^2 \div 3 = 1x^2$	$-21 \div 3 = -7$	$3(x^2 - 7)$
13.	$32x^2 + 24x + 16$	GCF = 8	$8(4x^2 + 3x + 2)$		
14.	$15x^3 + 15x^2$	GCF = $15x^2$	$15x^3 \div 15x^2 = 1x$	$15x^2 \div 15x^2 = 1$	$15x^2(x + 1)$
15.	$50x^3 - 25x$	GCF = $25x$	$50x^3 \div 25x = 2x^2$	$-25x \div 25x = -1$	$25x(2x^2 - 1)$

Factor by GCF (PAGE 9, PART 2)

1.	$7x^3 + 49x$	$7x(x^2 + 7)$
2.	$8x^{20} - 16x^{10}$	$8x^{10}(x^{10} - 2)$
3.	$6x + 36xy^3$	$6x(1 + 6y^3)$
4.	$15x^6 - 10x^2$	$5x^2(3x^4 - 2)$
5.	$33x^3 + 11x^2$	$11x^2(3x + 1)$
6.	$8x^{14} - 14x^4y$	$2x^4(4x^{10} - 7y)$
7.	$50x^3 - 25x$	$25x(2x^2 - 1)$
8.	$12x^8y^2 + 28xy^2$	$4xy^2(3x^7 + 7)$
9.	$10x^5 - 20x^4$	$10x^4(x - 2)$
10.	$300a^2b^3 + 150a^3b^2$	$150a^2b^2(2b + a$
11.	$9x^5y^2 - 12x^2y$	$3x^2y(3x^3y - 4)$
12.	$x^8y^2 - 21x^4y^4$	$x^4y^2(x^4 - 21y^2)$
13.	$3x^9 + 24x^7 + 18x^3$	$3x^3(x^6 + 8x^4 + 6$
14.	$2y^6 - 4y^5 + 4y$	$2y(y^5 - 2y^4 + 2)$
15.	$7x^3 + 14x^2 - 35x$	$7x(x^2 + 2x - 5)$

Factoring Trinomials A = 1 (PAGE 11)

1. $x^2 + 11x + 28$

$a = 1, b = 11, c = 28$

$a \cdot c = 1 \cdot 28 = 28$, $b = 11$

Factors of **28** that combine to equal **11**:

7 and 4; $7 \cdot 4 = 28$ ✓ and $7 + 4 = 11$ ✓

Answer: $(x + 7)(x + 4)$

2. $x^2 + 4x - 12$

$a = 1, b = 4, c = -12$

$a \cdot c = 1 \cdot (-12) = -12$, $b = 4$

Factors of **-12** that combine to equal **4**:

6 and (-2); $6 \cdot (-2) = -12$ ✓ and $6 + (-2) = 4$ ✓

Answer: $(x + 6)(x - 2)$

3. $x^2 + 13x + 30$

$a = 1, b = 13, c = 30$

$a \cdot c = 1 \cdot 30 = 30$, $b = 13$

Factors of **30** that combine to equal **13**:

10 and 3; $10 \cdot 3 = 30$ ✓ and $10 + 3 = 13$ ✓

Answer: $(x + 10)(x + 3)$

4. $x^2 - 10x - 11$

$a = 1, b = -10, c = -11$

$a \cdot c = 1 \cdot (-11) = -11$, $b = -10$

Factors of **-11** that combine to equal **10**:

(-11) and 1; $(-11) \cdot 1 = -11$ ✓ and $(-11) + 1 = -10$ ✓

Answer: $(x - 11)(x + 1)$

5. $x^2 - 12x + 35$

$a = 1, b = (-12), c = 35$

$a \cdot c = 1 \cdot 35 = 35$, $b = -12$

Factors of **35** that combine to equal **-12**:

(-7) and (-5); $(-7) \cdot (-5) = +35$ ✓ and $(-7) + (-5) = -12$ ✓

Answer: $(x - 7)(x - 5)$

6. $x^2 + 10x + 16$

$a = 1, b = 10, c = 16$

$a \cdot c = 1 \cdot 16 = 16$, $b = 10$

Factors of **16** that combine to equal **10**:

8 and 2; $8 \cdot 2 = 16$ ✓ and $8 + 2 = 10$ ✓

Answer: $(x + 8)(x + 2)$

7. $a^2 + 6x + 8$

$a = 1, b = 6, c = 8$

$a \cdot c = 1 \cdot 8 = 8$, $b = 6$

Factors of **8** that combine to equal **6**:

4 and 2; $4 \cdot 2 = 8$ ✓ and $4 + 2 = 6$ ✓

Answer: $(x + 4)(x + 2)$

8. $c^2 - c - 20$

$a = 1, b = -1, c = -20$

$a \cdot c = 1 \cdot (-20) = -20$, $b = (-1)$

Factors of **-20** that combine to equal **-1**:

(-5) and 4; $(-5) \cdot 4 = -20$ ✓ and $(-5) + 4 = -1$ ✓

Answer: $(x - 5)(x + 4)$

9. $y^2 + 16y + 63$

$a = 1, b = 16, c = 63$

$a \cdot c = 1 \cdot 63 = 63$, $b = 16$

Factors of **63** that combine to equal **16**:

9 and 7; $9 \cdot 7 = 63$ ✓ and $9 + 7 = 16$ ✓

Answer: $(x + 9)(x + 7)$

1. $6x^2 - 19x + 15$

$a \bullet c = 6 \bullet 15 = 90$, $b = -19$

Factors of **90** that combine to equal **-19**:

(-10) and (-9); $(-10) \bullet (-9) = +90$ ✓ **and** $(-10) + (-9) = -19$ ✓

$\left(x - \frac{10}{6}\right)\left(x - \frac{9}{6}\right) \Rightarrow\Rightarrow\Rightarrow$ reduce $\left(x - \frac{5}{3}\right)\left(x - \frac{3}{2}\right) \Rightarrow\Rightarrow\Rightarrow$ bring up $(3x - 5)(2x - 3)$

Answer: $(3x - 5)(2x - 3)$

2. $4y^2 - 11y + 6$

$a \bullet c = 4 \bullet 6 = 24$, $b = -11$

Factors of **24** that combine to equal **-11**:

(-8) and (-3); $(-8) \bullet (-3) = 24$ ✓ **and** $(-8) + (-3) = -11$ ✓

$\left(y - \frac{8}{4}\right)\left(y - \frac{3}{4}\right) \Rightarrow \left(y - 2\right)\left(y - \frac{3}{4}\right) \Rightarrow (y - 2)(4y - 3)$

Answer: $(y - 2)(4y - 3)$

3. $2a^2 + a - 3$

$a \bullet c = 2 \bullet -3 = -6$, $b = 1$

Factors of **-6** that combine to equal **1**:

3 and (-2); $3 \bullet (-2) = -6$ ✓ **and** $3 + (-2) = 1$ ✓

$\left(a + \frac{3}{2}\right)\left(a - \frac{2}{2}\right) \Rightarrow \left(a + \frac{3}{2}\right)(a - 1) \Rightarrow (2a + 3)(a - 1)$

Answer: $(2a + 3)(a - 1)$

4. $2x^2 + 3x - 9$

$a \bullet c = 2 \bullet -9 = -18$, $b = 3$

Factors of **-18** that combine to equal **3**:

6 and (-3); $6 \bullet (-3) = -18$ ✓ **and** $6 + (-3) = 3$ ✓

$\left(x + \frac{6}{2}\right)\left(x - \frac{3}{2}\right) \Rightarrow\Rightarrow\Rightarrow$ reduce $(x + 3)\left(x - \frac{3}{2}\right) \Rightarrow\Rightarrow\Rightarrow$ bring up $(x + 3)(2x - 3)$

Answer: $(x + 3)(2x - 3)$

5. $3c^2 - 8c + 4$

$a \bullet c = 3 \bullet 4 = 12$, $b = -8$

Factors of **12** that combine to equal **-8**:

(-6) and (-2); $(-6) \bullet (-2) = +12$ ✓ **and** $(-6) + (-2) = -8$ ✓

$\left(c - \frac{6}{3}\right)\left(c - \frac{2}{3}\right) \Rightarrow (c - 2)\left(c - \frac{2}{3}\right) \Rightarrow (c - 2)(3c - 2)$

Answer: $(c - 2)(3c - 2)$

6. $5x^2 + 19x + 12$

$a \bullet c = 5 \bullet 12 = 60$, $b = 19$

Factors of **60** that combine to equal **19**:

15 and 4; $15 \bullet 4 = 60$ ✓ **and** $15 + 4 = 19$ ✓

$\left(x + \frac{15}{5}\right)\left(x + \frac{4}{5}\right) \Rightarrow (x + 3)\left(x + \frac{4}{5}\right) \Rightarrow (x + 3)(5x + 4)$

Answer: $(x + 3)(5x + 4)$

7. $5a^2 - 18a + 9$

$a \bullet c = 5 \bullet 9 = 45$, $b = -18$

Factors of **45** that combine to equal **-18**:

(-15) and (-3); $(-15) \bullet (-3) = +45$ ✓ **and** $(-15) + (-3) = -18$ ✓

$\left(a - \frac{15}{5}\right)\left(a - \frac{3}{5}\right) \Rightarrow\Rightarrow\Rightarrow$ reduce $(a - 3)\left(a - \frac{3}{5}\right) \Rightarrow\Rightarrow\Rightarrow$ bring up $(a - 3)(5a - 3)$

Answer: $(a - 3)(5a - 3)$

8. $4b^2 - 15b - 25$

$a \bullet c = 4 \bullet -25 = -100$, $b = -15$

Factors of **-100** that combine to equal **-15**:

(-20) and 5; $(-20) \bullet 5 = -100$ ✓ **and** $(-20) + 5 = -15$ ✓

$\left(b - \frac{20}{4}\right)\left(b + \frac{5}{4}\right) \Rightarrow (b - 5)\left(b + \frac{5}{4}\right) \Rightarrow (b - 5)(4b + 5)$

Answer: $(b - 5)(4b + 5)$

9. $6y^2 + 37y + 6$

$a \bullet c = 6 \bullet 6 = 36$, $b = 37$

Factors of **36** that combine to equal **37**:

36 and 1; $36 \bullet 1 = 36$ ✓ **and** $36 + 1 = 37$ ✓

$\left(y + \frac{36}{6}\right)\left(y + \frac{1}{6}\right) \Rightarrow (y + 6)\left(y + \frac{1}{6}\right) \Rightarrow (y + 6)(6y + 1)$

Answer: $(y + 6)(6y + 1)$

ANSWERS

Factoring Trinomals A > 1 by GROUPING (PAGE 15)

ax^2 + factor(x) + factor(x) + c

The order of "factor(x) + factor(x)" is not important! The trinomial will factor by grouping perfectly either way you list the two factors in the middle.

1. $6x^2 - 19x + 15$

a•c = 6•15 = 90, b = -19

Factors of **90** that combine to equal **-19**:

(-10) and (-9); (-10)•(-9) = +90 ✓ and (-10) + (-9) = -19 ✓

ax^2 + factor(x) + factor(x) + c

⇨ $6x^2 - 10x \quad - 9x + 15$

GCF = 2x GCF = - 3

2x(3x - 5) - 3(3x - 5) ⇨ (2x - 3)(3x - 5)

Answer: (2x - 3)(3x - 5)

2. $4y^2 - 11x + 6$

a•c = 4•6 = 24, b = -11

Factors of **24** that combine to equal **-11**:

(-8) and (-3); (-8)•(-3) = 24 ✓ and (-8) + (-3) = -11 ✓

ax^2 + factor(x) + factor(x) + c

⇨ $4y^2 - 8y \quad - 3y + 6$

GCF = 4y GCF = - 3

4y(y - 2) - 3(y - 2) ⇨ (4y - 3)(y - 2)

Answer: (4y - 3) (y - 2)

3. $2a^2 + a - 3$

a•c = 2•-3 = -6, b = 1

Factors of **-6** that combine to equal **1**:

3 and (-2); 3•(-2) = -6 ✓ and 3 + (-2) = 1 ✓

ax^2 + factor(x) + factor(x) + c

⇨ $2a^2 + 3a \quad - 2a - 3$

GCF = a GCF = - 1

a(2a + 3) - 1(2a + 3) ⇨ (a - 1)(2a + 3)

Answer: (a - 1) (2a + 3)

4. $2x^2 + 3x - 9$

a•c = 2•-9 = -18, b = 3

Factors of **-18** that combine to equal **3**:

6 and (-3); 6•(-3) = -18 ✓ and 6 + (-3) = 3 ✓

ax^2 + factor(x) + factor(x) + c

⇨ $2x^2 + 6x \quad - 3x - 9$

GCF = 2x GCF = - 3

2x(x + 3) - 3(x + 3) ⇨ (2x - 3)(x + 3)

Answer: (2x - 3)(x + 3)

5. $3c^2 - 8c + 4$

a•c = 3•4 = 12, b = -8

Factors of **12** that combine to equal **-8**:

(-6) and (-2); (-6)•(-2) = +12 ✓ and (-6) + (-2) = -8 ✓

ax^2 + factor(x) + factor(x) + c

⇨ $3c^2 - 6c \quad - 2c + 4$

GCF = 3c GCF = - 2

3c(c - 2) - 2(c - 2) ⇨ (3c - 2)(c - 2)

Answer: (3c - 2)(c - 2)

6. $5x^2 + 19x + 12$

a•c = 5•12 = 60, b = 19

Factors of **60** that combine to equal **19**:

15 and 4; 15•4 = 60 ✓ and 15 + 4 = 19 ✓

ax^2 + factor(x) + factor(x) + c

⇨ $5x^2 + 4x + 15x + 12$ (switched my typical order of factors here)

GCF = x GCF = 3

x(5x + 4) + 3(5x + 4) ⇨ (x + 3)(5x + 4)

Answer: (5x + 4)(x + 3)

7. $5a^2 - 18a + 9$

a•c = 5•9 = 45, b = -18

Factors of **45** that combine to equal **-18**:

(-15) and (-3); (-15)•(-3) = +45 ✓ and (-15)+ (-3) = -18 ✓

ax^2 + factor(x) + factor(x) + c

⇨ $5a^2 - 15a \quad - 3a + 9$

GCF = 5a GCF = - 3

5a(a - 3) - 3(a - 3) ⇨ (5a - 3)(a - 3)

Answer: (5a - 3)(a - 3)

8. $4b^2 - 15b - 25$

a•c = 4•-25 = -100, b = -15

Factors of **-100** that combine to equal **-15**:

(-20) and 5; (-20)•5 = -100 ✓ and (-20) + 5= -15 ✓

ax^2 + factor(x) + factor(x) + c

⇨ $4b^2 - 20b + 5b - 25$

GCF = 4b GCF = - 3

4b(b - 5) + 5(b - 5) ⇨ (4b + 5)(b - 5)

Answer: (4b + 5) (b - 5)

9. $6y^2 + 37y + 6$

a•c = 6•6 = 36, b = 37

Factors of **36** that combine to equal **37**:

36 and 1; 36•1 = 36 ✓ and 36 + 1 = 37 ✓

ax^2 + factor(x) + factor(x) + c

⇨ $6y^2 + 36y \quad + y + 6$

GCF = 6y GCF = 1

6y(y + 6) + 1(y + 6) ⇨ (6y + 1)(y + 6)

Answer: (6y + 1)(y + 6)

Factor Sum and Difference of Squares PART 1 (PAGE 17)

1.	$4x^2 - 9$	2x and 3	$(2x + 3)(2x - 3)$		
2.	$100x^2 - 81$	10x and 9	$(10x + 9)(10x - 9)$		
3.	$25x^2 + 9$	PRIME			
4.	$a^2 - 121$	a and 11	$(a + 11)(a - 11)$		
5.	$9y^2 - 4z^2$	3y and 2z	$(3y + 2z)(3y - 2z)$		
6.	$144 - x^2$	12 and x	$(12 + x)(12 - x)$		
7.	$49a^2 - 36b^2$	7a and 6b	$(7a + 6b)(7a - 6b)$		
8.	$64b^2 + 9$	PRIME			
9.	$64b^2 - 9$	8b and 3	$(8b + 3)(8b - 3)$		
10.	$8x^2 - 50$	GCF: 2	$2(4x^2 - 25)$	2x and 5	$2(2x + 5)(2x - 5)$
11.	$45y^2 - 125$	GCF: 5	$5(9y^2 - 25)$	3y and 5	$5(3y + 5)(3y - 5)$
12.	$3x^2 - 27$	GCF: 3	$3(x^2 - 9)$	x and 3	$3(x + 3)(x - 3)$
13.	$(x + y)^2 - 9$	$(x + y)$ and 3	$[(x + y) + 3][(x + y) - 3]$		
14.	$(7a - 4b)^2 - 25$	$(7a - 4b)$ and 5	$[(7a - 4b) + 5][(7a - 4b) - 5]$		
15.	$(11x + 6y)^2 - 16$	$(11x + 6y)$ and 4	$[(11x + 6y) + 4][(11x + 6y) - 4]$		

Factor Sum and Difference of Squares PART 2 (PAGE 17)

1.	$16x^2 - 1$	4x and 1	$(4x + 1)(4x - 1)$
2.	$121x^2 - 64$	11x and 8	$(11x + 8)(11x - 8)$
3.	$4a^2 - 9b^2$	2a and 3b	$(2a + 3b)(2a - 3b)$
4.	$49a^2 - 121$	7a and 11	$(7a + 11)(7a - 11)$
5.	$144y^2 - z^2$	12y and z	$(12y + z)(12y - z)$
6.	$400 - x^2$	20 and x	$(20 + x)(20 - x)$
7.	$36x^2 + 9$	PRIME	
8.	$625b^2 + 9$	PRIME	
9.	$625b^2 - 9$	25b and 3	$(25b + 3)(25b - 3)$
10.	$100x^2 - 81$	10x and 9	$(10x + 9)(10x - 9)$
11.	$49y^2 - 121$	7y and 11	$(7y + 11)(7y - 11)$
12.	$x^2 - 144$	x and 12	$(x + 12)(x - 12)$
13.	$(5x + 2y)^2 - 16$	$(5x + 2y)$ and 4	$[(5x + 2y) + 4][(5x + 2y) - 4]$
14.	$(a - 4b)^2 - 9$	$(a - 4b)$ and 3	$[(a - 4b) + 3][(a - 4b) - 3]$
15.	$(3x + 4y)^2 - 49$	$(3x + 4y)$ and 7	$[(3x + 4y) + 7][(3x + 4y) - 7]$

Factor Sum and Difference of Cubes (PAGE 19)

SUM $a^3 + b^3 = (a + b)(a^2 - ab + b^2)$ DIFFERENCE $a^3 - b^3 = (a - b)(a^2 + ab + b^2)$

1.	$x^3 - 27$	cube roots: x and 3	-	SOAP	$(x - 3)(x^2 + 3x + 9)$
2.	$1000y^3 - 8$	cube roots: 10y and 2	-	SOAP	$(10y - 2)(100y^2 + 20y + 4)$
3.	$125a^3 + 343$	cube roots: 5a and 7	+	SOAP	$(5a + 7)(25a^2 - 35a + 49)$
4.	$c^3 - 64$	cube roots: c and 4	-	SOAP	$(c - 4)(c^2 + 4c + 16)$
5.	$8y^3 + z^3$	cube roots: 2y and z	+	SOAP	$(2y + z)(4y^2 - 2yz + z^2)$
6.	$x^3 - 512$	cube roots: x and 8	-	SOAP	$(x - 8)(x^2 + 8x + 64)$
7.	$343a^3 - 27b^3$	cube roots: 7a and 3b	-	SOAP	$(7a - 3b)(49a^2 + 21ab + 9b^2)$
8.	$p^3 + 125$	cube roots: p and 5	+	SOAP	$(p + 5)(p^2 - 5p + 25)$
9.	$64x^3 + 729$	cube roots: 4x and 9	+	SOAP	$(4x + 9)(16x^2 - 36x + 81)$
10.	$27x^3 - 125$	cube roots: 3x and 5	-	SOAP	$(3x - 5)(9x^2 + 15x + 25)$
11.	$512y^3 + 125$	cube roots: 8y and 5	+	SOAP	$(8y + 5)(64y^2 - 40y + 25)$
12.	$343x^3 - 27y^3$	cube roots: 7x and 3y	-	SOAP	$(7x - 3y)(49x^2 + 21xy + 9y^2)$
13.	$a^3 - 64b^3$	cube roots: a and 4b	-	SOAP	$(a - 4b)(a^2 + 4ab + 16b^2)$
14.	$729x^3 + 125y^3$	cube roots: 9x and 5y	+	SOAP	$(9x + 5y)(81x^2 - 45xy + 25y^2)$
15.	$250t^3 - 16r^3$ **GCF: 2** $2(125t^3 - 8r^3)$	cube roots: 5t and 2r	-	SOAP	$2(5t - 2r)(25t^2 + 10rt + 4r^2)$

Factor Special Case (binomial squared) (PAGE 21)

1.	$x^2 + 10x + 25$	x and 5	$(x + 5)^2$
2.	$x^2 - 10x + 25$	x and -5	$(x - 5)^2$
3.	$y^2 + 12x + 36$	y and 6	$(y + 6)^2$
4.	$y^2 - 12y + 36$	y and -6	$(y - 6)^2$
5.	$16a^2 - 40a + 25$	4a and -5	$(4a - 5)^2$
6.	$16a^2 + 40a + 25$	4a and 5	$(4a + 5)^2$
7.	$x^2 - 6xy + 9y^2$	x and -3y	$(4a + 5)^2$
8.	$x^2 + 6xy + 9y^2$	x and 3y	$20y + 4)^2$
9.	$9x^2 - 48xy + 64y^2$	3x and -8y	$20y + 4)^2$
10.	$9x^2 + 48xy + 64y^2$	3x and 8y	$20y + 4)^2$
11.	$4a^2 + 8ab + 4b^2$	2a and 2b	$20y + 4)^2$
12.	$x^2 - 2xy + y^2$	x and -y	$20y + 4)^2$
13.	$25x^2 + 70xy + 49y^2$	5x and 7y	$20y + 4)^2$
14.	$2x^2 + 20x + 50$	**GCF $2(x^2 + 10x + 25)$**	
		x and 5	$2(x + 5)^2$
15.	$8x^2 + 16xy + 8y^2$	**GCF $8(x^2 + 2x + y)$**	
		x and y	$8(x + y)^2$

ANSWERS

Review Quiz PAGE 23

Find the GCF of the following:

1. *a.* 15 and 25 = 5 *b.* 8 and 34 = 2

2. *a.* $12x^2$ and $8x$ = $4x$ *b.* y^2 and $6z$ = 1

Factor by GCF:

3. *a.* $8x - 18$ *b.* $25a^2 + 10a$

a. $2(4x - 9)$ *b.* $5a(5a + 2)$

c. $5b^2 + 35b$ *d.* $45x^3 + 9x^2 + 9x$

c. $5b(b + 7)$ *d.* $9x(5x^2 + x + 1)$

Factor:

4. $a^2 - a - 20$ $(a + 4)(a - 5)$
5. $x^2 + 6x + 8$ $(x + 4)(x + 2)$
6. $y^2 + 2y + 9$ PRIME
7. $n^2 - 10n + 25$ $(n - 5)^2$
8. $x^2 - 7x - 30$ $(x - 10)(x + 3)$
9. $2a^2 - 4a - 70$ GCF 1st: 2

$2(a^2 - 2a - 35) = 2(a - 7)(a + 5)$

10. $y^2 - 11y + 28$ $(y - 7)(y - 4)$
11. $x^2 + 14x + 40$ $(x + 10)(x + 4)$
12. $x^2 + 14x + 45$ $(x + 9)(x + 5)$
13. $z^2 - z - 56$ $(z - 8)(z + 7)$
14. $a^2 - 81$ $(a + 9)(a - 9)$
15. $5x^2 - 18x + 9$ $(5x - 3)(x - 3)$
16. $8a^3 - 27$

$(2a - 3)(4a^2 + 6a + 9)$

17. $3n^2 + 8n - 5$ PRIME
18. $64x^2 - 49y^2$ $(8x + 7y)(8x - 7y)$
19. $3a^2 - 17a - 20$ $(3a - 20)(a + 1)$
20. $7y^2 + 50y + 7$ $(7y + 1)(y + 7)$
21. $36x^2 + 9$ PRIME
22. $(5x + 6y)^2 - 49$

$(5x + 6y + 7)(5x + 6y - 7)$

23. $81a^2 - 100b^2$ $(9a + 10b)(9a - 10b)$
24. $2x^3 + 16y^3$ GCF: 2

$2(x^3 + 8y^3) = 2(x + 2y)(x^2 - 2xy + 4y^2)$

25. $9y^2 - 30yz + 25z^2$ $(3y - 5z)^2$
26. $8x^2 - 12x - 8$ GCF: 4

$4(2x^2 - 3x - 2) = 4(2x + 1)(x - 2)$

27. $a^2 + 2ab + b^2$ $(a + b)^2$
28. $27z^3 - 343$ $(3z - 7)(9z^2 + 21z + 49)$
29. $5c^2 - 11x - 12$ $(c - 3)(5c + 4)$
30. $25x^2 + 30x + 9$ $(5x + 3)^2$

Extra Practice PAGE 25

1. $5x^3 + 40y^3 =$ $\mathbf{5(x + 2y)(x^2 - 2xy + 4y^2)}$
2. $16a^2 + 56a + 49 =$ $\mathbf{(4a + 7)^2}$
3. $2x^2 - 16x + 32 =$ $\mathbf{2(x - 4)^2}$
4. $8y^2 - 16 - 28y =$ $\mathbf{4(2y + 1)(y - 4)}$
5. $25 - 10b + b^2 =$ $\mathbf{(b - 5)^2}$ **or** $\mathbf{(5 - b)^2}$ **same**
6. $6a^6 + a^3 - 2 =$ $\mathbf{(3a^3 + 2)(2a^3 - 1)}$
7. $81 - z^4 =$ $\mathbf{(9 + z^2)(3 + z)(3 - z)}$
8. $(x + y)^2 - 25 =$ $\mathbf{(x + y + 5)(x + y - 5)}$
9. $y^2 - 144 =$ $\mathbf{(y + 12)(y - 12)}$
10. $25 - a^2 =$ $\mathbf{(5 + a)(5 - a)}$
11. $5x^2y^3 - 15x^3y^2 =$ $\mathbf{5x^2y^2(y - 3x)}$
12. $x^2 + 9x + 20 =$ $\mathbf{(x + 5)(x + 4)}$
13. $c^3 + 9c^2 =$ $\mathbf{c^2(c + 9)}$
14. $y^3 + 8 =$ $\mathbf{(y + 2)(y^2 - 2y + 4)}$
15. $64x^4 + x =$ $\mathbf{x(4x + 1)(16x^2 - 4x + 1)}$
16. $r^6 - 64 =$ $\mathbf{(r + 2)(r^2 - 2r + 4)(r - 2)(r^2 + 2r + 4)}$
17. $6y^2 + 23y + 20 =$ $\mathbf{(3y + 4)(2y + 5)}$
18. $8r^2 - 6r - 9 =$ $\mathbf{(4r + 3)(2r - 3)}$
19. $3x + x^2 - 10 =$ $\mathbf{(x + 5)(x - 2)}$
20. $a^2 + 5a - 84 =$ $\mathbf{(a + 12)(a - 7)}$
21. $36 - (x - y)^2 =$ $\mathbf{(6 + x - y)(6 - x + y)}$
22. $9a^3 - 49a =$ $\mathbf{a(3a + 7)(3a - 7)}$
23. $6x^2 - 7x - 10 =$ $\mathbf{(6x + 5)(x - 2)}$
24. $(1/4) - x^2 =$ $\mathbf{[(1/2) + x][(1/2) - x]}$
25. $64x^3 + 27 =$ $\mathbf{(4x + 3)(16x^2 - 12x + 9)}$
26. $81 - 18a + a^2 =$ $\mathbf{(9 - a)^2}$ **or** $\mathbf{(a - 9)^2}$
27. $y^2 - 12y + 36 =$ $\mathbf{(y - 6)^2}$
28. $121z^2 - 1 =$ $\mathbf{(11z + 1)(11z - 1)}$
29. $6c^2 + 12c + 6 =$ $\mathbf{6(c + 1)^2}$
30. $25x^2 + 36y^2 =$ **PRIME**
31. $t^2 - 8t - 48 =$ $\mathbf{(t - 12)(t + 4)}$
32. $11a^2 - 11b^2 =$ $\mathbf{11(a + b)(a - b)}$
33. $3q^5 - 12q^3 =$ $\mathbf{3q^3(q + 2)(q - 2)}$
34. $8xy^4 + 8x^4y =$ $\mathbf{8xy(y + x)(y^2 - xy + x^2)}$
35. $5x^2 - 2x + 3 =$ **PRIME**
36. $y^3 - 343 =$ $\mathbf{(y - 7)(y^2 + 7y + 49)}$
37. $36x^2 - 9 =$ $\mathbf{9(2x + 1)(2x - 1)}$
38. $a^2 + 5a - 36 =$ $\mathbf{(a - 4)(a + 9)}$
39. $216 - x^3 =$ $\mathbf{(6 - x)(36 + 6x + x^2)}$
40. $9c^5 + 99c^3d^5 =$ $\mathbf{9c^3(c^2 + 11d^5)}$
41. $f^2 - 5f - 14 =$ $\mathbf{(f - 7)(f + 2)}$
42. $9x^2y^2 - 25y^4 =$ $\mathbf{y^2(3x + y)(3x - y)}$
43. $s^2 - 3s - 2 =$ **PRIME**
44. $6y^3 + 48 =$ $\mathbf{6(y + 2)(y^2 - 2y + 4)}$
45. $3x^2 - 34x - 24 =$ $\mathbf{(3y + 2)(y - 12)}$
46. $x^2 + 8x + 16 =$ $\mathbf{(x + 4)^2}$
47. $x^6 - 1 = (x^3+1)(x^3-1) =$ $\mathbf{(x+1)(x^2-x+1)(x-1)(x^2+x+1)}$
48. $27a^2 - 30a - 8 =$ $\mathbf{(9a + 2)(3a - 4)}$

FACTORING REFERENCE by April Chloe Terrazas

IG @bestmathtutor www.aprilisthebomb.com

$x^2 + 10x + 16 = (x + 8)(x + 2)$ $x^2 + 13x + 30 = (x + 10)(x + 3)$ $x^2 + 4x - 12 = (x + 6)(x - 2)$ $x^2 + 7x + 12 = (x + 5)(x + 2)$ $x^2 - 5x + 6 = (x - 3)(x - 2)$

Graphing Calculator Tip - FACTOR PAIRS

Go to **Y=**

Enter the number you want factors of **÷ x**

For example: *$y = 36/x$ $y = -125/x$*

Go to **TABLE**

View all factor pairs of the number! ⇨

ex: 1, 250; 2, 125; 5, 50...

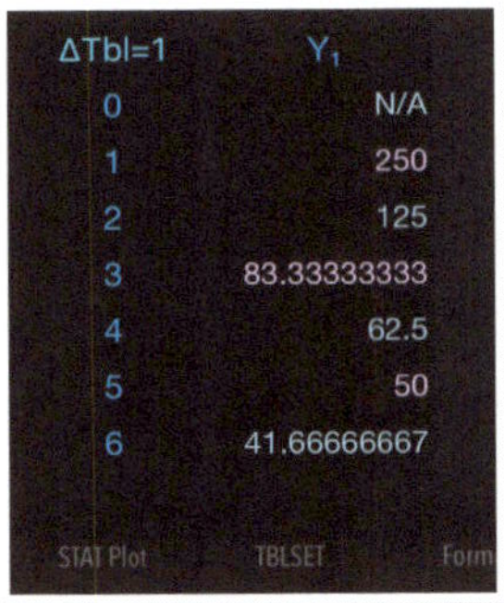

This is a phone app called ***Graphing Calculator Plus*** *available in the App Store.*

Factor by GCF

Factor $12x^2 + 3x$ **GCF = 3x**

Divide the GCF from all terms

$12x^2 \div 3x = 4x$ and $3x \div 3x = 1$

(or, think of what term you multiply by the GCF to get the term back when you distribute to check your answer)

or $3x \bullet 4x = 12x^2$ and $3x \bullet 1 = 3x$

answer: $3x(4x + 1)$

Examples

$15x^2 + 3x = 3x(5x + 1)$

$45x^3y^5 + 20x^2y^3 = 5x^2y^3(9xy^2 + 4)$

$7a^4b^2c^4 - 14a^2b^2c^5 = 7a^2b^2c^4(a^2 - 2c)$

$6x^5 + 3x^3 + x = x(6x^4 + 3x^2 + 1)$

Factoring Trinomials $ax^2 + bx + c$

Step 1: Multiply a•c and find factors of that term that combine to equal b. **Step 2**: If a=1, simply write those factors in a pair with parentheses. Ex: $(x + 2)(x + 3)$ If a>1, factor by grouping or divide each factor by a.

Factor $3x^2 - 17x - 20$

GROUPING

Step 1: Factors of -60 that combine to equal -17: ***3 and -20***.

Step 2: Rewrite the polynomial starting with the same "a" value, using the two factors as middle terms, and ending with the "c" value: $3x^2 + 3x - 20x + 12$.

Step 3: GROUP the first two and the second two terms, find GCF:

$3x^2 + 3x$ $- 20x - 20$

GCF: **x**   **-20**

Divide GCF out of both terms:

$3x(x + 1)$ $-20(x + 1)$

Step 4: Combine the GCF's as one binomial **(3x - 20)** next to the other binomial $(x + 1)$ to complete factoring $(3x - 20)(x + 1)$.

Step 5: FOIL and check your work.

DIVIDE BY A

Same Step 1

Step 2: In the parentheses, x + one factor, and x + the other factor.

⇨ $(x+3)(x - 20)$.

Step 3: Divide each factor by the "a" value, **3**.

$(x + 3/3)(x - 20/3)$

Step 4:

If the fraction does NOT simplify, bring the denominator UP in front of the x ⇨ $(x+1)(3x - 20)$.

Step 5: FOIL and check your work.

Sum and Difference of Squares

NO Sums of squares are ALWAYS PRIME **NO**

Sums of squares **cannot be factored**

$x^2 + y^2$ $9x^2 + 36y^2$ $121a^2 + 1$ $a^2 + b^2$

Difference of Squares $a^2 - b^2 = (a + b)(a - b)$

Step 1: Take the square root of "a" and "b"

Step 2: Plug those values into the template

PLUS and MINUS the exact SAME VALUE

$a^2 - 4 = (a + 2)(a - 2)$ $y^2 - 9 = (y + 3)(y - 3)$

$x^2 - 16 = (x + 4)(x - 4)$ $36 - b^2 = (6 + b)(6 - b)$

$49x^2 - 25y^2 = (7a + 5y)(7a - 5y)$

$49x^4 - 25y^4 = (7x^2 + 5y^2)(7x^2 - 5y^2)$

$(a + b)^2 - 25 = (a + b + 5)(a + b - 5)$

$x^4 - 16 = (x^2 + 4)(x^2 - 4) = (x^2 + 4)(x + 2)(x - 2)$

Binomial Squared

Keep an eye out for patterns of squares!

$a^2 + 2ab + c^2 = (a + b)^2$

$a^2 - 2ab + c^2 = (a - b)^2$

You will begin to recognize squared binomials easily.

Know the squares of 2^2 - 10^2

The **1st and 3rd term** are ALWAYS SQUARES^{222}

The **3rd** term is ALWAYS POSITIVE+++

The **middle** term = 2•ab

$x^2 + 16x + 64 = (x + 8)^2$

$x^2 - 16x + 64 = (x - 8)^2$

$9a^2 + 12ab + 4b^2 = (3a + 2b)^2$

$9a^2 - 12ab + 4b^2 = (3a - 2b)^2$

$4x^2 + 36x + 81 = (2x + 9)^2$

$4x^2 - 36x + 81 = (2x - 9)^2$

Step 1: Take the square root of the a and c term. **Step 2**: Those roots are the a and b terms that go in parentheses $(a + b)^2$ or $(a - b)^2$. The sign on b term in the factored binomial is the same as the middle term in the trinomial. *Ex:* $x^2 + 16x + 64$. Square root of a and c equals x and 8 respectively. The middle term of trinomial is (+) so b term in factored binomial answer will also be (+): $(x + 8)^2$

Sum and Difference of Cubes $a^3 + b^3$ $a^3 - b^3$

Step 1: Take the cube root of "a" and "b"

Step 2: Input using **SOAP** (same, opposite, always positive)

Sum $a^3 + b^3 = (a + b)(a^2 - ab + b^2)$

$x^3 + 27 = (x + 3)(x^2 - 3x + 9)$

$125a^3 + 343b^3 = (5a + 7b)(25a^2 - 35ab + b^2)$

$729x^2 + 1 = (9x + 1)(81x^2 - 9x + 1)$

Difference $a^3 - b^3 = (a - b)(a^2 + ab + b^2)$

$x^3 - 27 = (x - 3)(x^2 + 3x + 9)$

$125a^3 - 343b^3 = (5a - 7b)(25a^2 + 35ab + b^2)$

$729x^2 - 1 = (9x - 1)(81x^2 + 9x + 1)$

Important stuff to remember